Pencil &
Paper
Games

Pencil & Paper Games

Karl-Heinz Koch

Sterling Publishing Co., Inc.   New York

Translated by Elisabeth E. Reinersmann

Library of Congress Cataloging-in-Publication Data

Koch, Karl-Heinz.
    [Spiele mit Papier und Bleistift. English]
    Pencil & paper games / by Karl-Heinz Koch.
      p.   cm.
    Translation of: Spiele mit Papier und Bleistift.
    Includes index.
    ISBN 0-8069-8262-4
      1. Pencil games. 2. Group games. 3. Mathematical recreations.
  I. Title. II. Title: Paper and pencil games. III. Title: Spiele
mit Papier und Bleistift.
    GV1493.K63   1991
    793.73—dc20   90-24245
                                               CIP

10 9 8 7 6 5 4 3 2 1

First paperback edition published in 1992 by
Sterling Publishing Company, Inc.
387 Park Avenue South, New York, N.Y. 10016
English translation © 1991 by Sterling Publishing Company
Originally published in German under the title
Spiele mit Papier und Bleistift: d. 55 Schonsten Schriebspiel
Distributed in Canada by Sterling Publishing
% Canadian Manda Group, P.O. Box 920, Station U
Toronto, Ontario, Canada M8Z 5P9
Distributed in Great Britain and Europe by Cassell PLC
Villiers House, 41/47 Strand, London WC2N 5JE, England
Distributed in Australia by Capricorn Link Ltd.
P.O. Box 665, Lane Cove, NSW 2066
Manufactured in the United States of America

Sterling ISBN 0-8069-8262-4 Trade
       ISBN 0-8069-8263-2 Paper

CONTENTS

PREFACE

Everybody enjoys pencil-and-paper (or writing) games, but traditionally they've been something particularly students have loved. For generations, students have fought exciting, dramatic battles—unobtrusively under their desks. Being in math class sometimes meant using graph paper more often for pencil-and-paper games than for assignments.

Among the pencil-and-paper games presented in this book, you can find games at all levels of difficulty, from easy games such as "Bingo" and "Lotto" to complicated games involving strategic battles fought between keenly competitive rivals. There are also games that call for mathematical calculations, and artwork games limited only by your imagination.

However, what all these games have in common is their basic simplicity. All that's really needed is knowing the rules and having a piece of (usually graphed) paper and some pencils (if possible, multicolored) on hand. No fancy, expensive playing board, no figures, no play money are needed.

Aside from glossy packaged versions of a few of these games that are available in stores, books such as this one are the only place where you can find writing games. Usually, however, these games are passed on by word of mouth, which unfortunately makes them very difficult to come upon.

In principle, every board game, from checkers to chess, can be played with paper and pencil. You transfer the configuration, including the appropriate figures, from the board to a sheet of paper. If a figure is moved, you erase it from where it is and mark its new position. However, games in which you move figures about this way cannot really be considered pencil-and-paper games. With pencil-and-paper games, every individual move made by each of the players is recorded on paper and remains there until the end of the game.

Also, during the course of a writing game, often a certain structure is created, not unlike those of other creative processes, such as writing a book, painting a picture, or coming up with a mathematical equation.

Looking into the origin of games, we can see that they can be divided into three major categories. The first category consists of games that are considered common property (such as the game known as "Battleship" in America and "Sink the Ship" in Germany), since they have been known and played for such a long time. The name of the person who first thought them up generally remains unknown. The many different versions of board and figure games, which entrepreneurs have attempted to market, represent the second category. I've made certain alterations in some of these games and thereby turned them into legitimate pencil-and-paper games. The third category of games consists of those that have been presented to the world by creative players or professional authors of games, and has come onto the scene rather recently. These games can be considered "mental constructions" and are exempt from copyright. This is a shame in certain respects; for instance, I would have received a copyright for my math game "Addition," and then every little store selling my game would owe me my fair share of the selling price. But for all those people who enjoy playing these games, it's a real gift.

It is because writing games are basically constructions perceived by the mind that they have become common property. However, this also means that I didn't have to limit myself in this book to only those games in the first category—those old and faithful companions—but could give the reader samples from the second and third categories as well.

Courtesy demands that the author of a game be given credit. But that is easier said than done, since, unlike people who write books, the names of people who invent games are very seldom known. When someone passes on the rules of a particular game, the inventor is rarely mentioned. Therefore, I offer a deep and apologetic bow to every author who has not been given credit in this book.

Many of these games are based on strategy, which I have divided into two categories: strategic games for two, and strategic games for more than two people. There are also action games where the pencil is actively employed. The puzzles are games for the solitary player. Both letter and number games are presented in separate chapters. The last category, cooperation games, are for those players who simply play for the fun of it and don't care about competition.

For quick orientation, I have included the degree of difficulty, from "1" to "6", under the title of each game. If the degree of difficulty is "1", the game is uncomplicated and can almost be played in your sleep; if it is "6", the game requires a high degree of attentiveness. Likewise, I have stated how many people can participate in playing each respective game.

But enough of all this talk—let the games begin: On your mark, pencils, write!

Strategic Games
for Two

SINK THE SHIP

Degree of difficulty: 3
Players: two

It's only fair to begin this book with "Sink the Ship," since it is *the* quintessential pencil-and-paper game and the model for many similar games.

The opposing vice admirals are each in command of 10 ships: four U-boats (occupying one square); three speedboats (2 squares), two destroyers (3 squares), and one battleship (4 squares). The ships of each player are located in an "ocean," consisting of 10×10 squares. The goal for each player is to send every one of the opponent's ships to their watery grave, while at least one of his own is still afloat.

Begin by taking two pieces of graph paper and outlining two fields of 10×10 squares on each: one for positioning your own fleet in the water, and the other for recording the damage—or lack thereof—you do to the fleet of your opponent. Assign the letters A to K (leaving out the letter J) to the vertical squares above the top row, and the numbers 1 to 10 to the horizontal squares on the left side, as shown in the illustration on page 16. You are now able to identify each one of the 100 squares: D-4, H-2, and so on. Sketch out your own fleet, using the specified number of squares for each ship (see the illustration).

Conceal your "ocean" from the eyes of your opponent, when you mark the position of your own flotilla. The ships can be positioned either horizontally or vertically and can occupy any of the squares on the outer perimeter, but they are not to touch each other, not even at the corners! It's important to

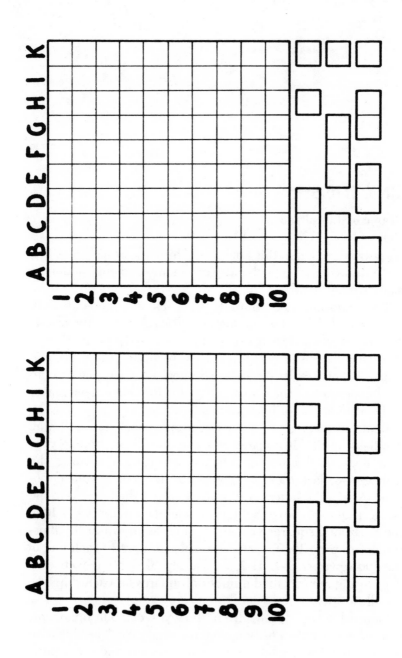

16

choose positions carefully because you aren't allowed to make any changes once the game has begun.

Since the squares surrounding each ship have to remain free, a battleship placed in the middle of the ocean takes up a lot of extra space—14 additional squares to be exact. However, if a battleship is positioned along the outside, only 8 more squares will be involved, and if pushed into a corner, it will take up only 6 additional squares. Either way leaves a considerable amount more "ocean space" for smaller boats to hide in. A truly tricky admiral will put as much as he can into the spaces along the perimeter. This will give him more room in which to maneuver his U-boats, which after all take up only one square. If your strategy, however, becomes too obvious, your opponent will be able to flush your ships out and send them to the bottom of the ocean in no time.

The battle begins. Taking turns, each player calls the coordinate of a square—let's say, F-8—and his opponent must reply by either answering with "Water," if indeed no ship has been hit, or with "Hit," if the square is occupied by a ship or part of a ship. In the latter case, the opponent is allowed to "fire another shot." The opponent is in an ideal situation: He can assume that one of the four squares surrounding the "hit"—F-7, F-9, E-8, or G-8—is occupied by the ship already damaged by his first shot.

As long as a player has made a "hit," he can continue playing. If the last square of a ship has been hit, the opponent must declare "Ship sinking." (For a U-boat, unfortunately, the first hit is also the last.) A player has one more shot after he has sunk a boat, regardless of how much is left of his own fleet.

Accurate records should be kept about the proceedings. The shots fired at the opponent are recorded in the right field. If a shot went into the water, mark the square with a dot. A square that *is* a ship, or part of one, is marked with a cross. For the sake of clarity, mark all squares surrounding the sunken ship with a dot also, since they cannot be occupied.

In the left field, record the fate of your own fleet, not just the shots that went into the water, but also those that were hits. This will give you a good view of your opponent's progress.

Since people enjoy diversity and like to invent new rules, this game exists in many different forms. Two in particular are favorites. In one variation, ships can also be positioned diagonally, meaning that the squares constituting a ship touch only

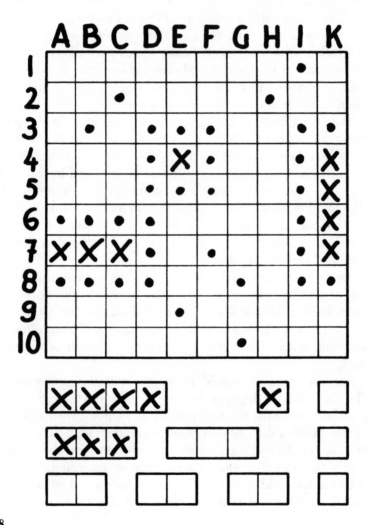

at the corners. This increases the difficulty in locating the rest of the squares in order to sink the ship. But in spite of that advantage, experienced admirals seldom use this strategy, since so much extra space is used up. A diagonally positioned battleship, for instance, is surrounded by 20 squares—squares that can't be used for other boats. A vertical or horizontal battleship, on the other hand, is only surrounded by 14 squares.

Another variation of the game is to expand the "ocean" to a field of 15 × 15 squares. The number of ships in the fleet is also increased to five U-boats, four speedboats, three destroyers, two battleships, and one aircraft carrier (taking up 5 squares). While this variation prolongs the game, it doesn't necessarily increase the excitement.

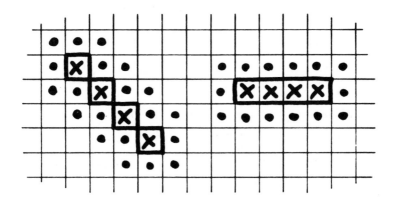

RADAR

Degree of difficulty: 6
Players: two

The fun I had with "Sink the Ship" was soon replaced by boredom due to the somewhat mindless "shoot 'em up" mentality of the game. But it served as an incentive for me to develop a more peaceful version, one that doesn't depend so much on luck and chance but on a sharp mind.

With "Radar," double the dimensions of the ocean and the ships used in "Sink the Ship." Thus, each player needs two fields, measuring 20×20. Identification of the squares is the same, with the letters on top and the numbers on the side. Likewise, double the size of each boat. Use 2×2 squares for four U-boats, 2×4 rectangles for the three speedboats, 2×6 rectangles for the two destroyers, and 2×8 rectangles for the battleship (a battleship is shown on page 21). They can only be positioned horizontally or vertically, and they cannot touch each other, not even at the corners.

The players, as in "Sink the Ship," take turns, but they do not shoot at each other—they only try to locate the position of their opponent's ships. With "Sink the Ship," you know the results immediately: either "water" or "hit." With this game, taking the bearing of an object is only approximate, and repeated bearings are necessary in order to get the information needed to identify and locate a ship.

A player, on his turn, names three connected squares, either in horizontal, vertical, or diagonal directions—for instance, A-1 to A-3; A-1 to C-1; A-1, B-2, C-3 or A-3, B-2, C-1. If all three squares are empty, your opponent answers "Water"; otherwise, his reply is "One fix," "Two fixes," or "Three fixes." The problem for the player is that with the replies "One fix" or "Two fixes," he doesn't know which one or two of the three squares is part of a ship. Additionally, if the opponent has the fix on two squares, they might belong to two different ships with a "water" square separating them. If a player gets a fix on two or more squares, he continues with his turn.

Keeping track of progress during this game becomes a bit

more complicated. When your opponent's response is "Water," mark the three specified squares with a dot, as before; likewise, mark three "fixes" with crosses. These results are unambiguous. For one or two fixes, I recommend a single or double line drawn through the "bearing" squares. The direction of the line would indicate the possible or probable connection between them (see the illustration). In contrast to the rule in "Sink the Ship," a player doesn't have to announce whether his opponent has a fix on a square that completes a ship. To the contrary, it is the player's responsibility to announce the exact location of a given ship of his opponent's, and that can be done at any time. It is not necessary to wait until

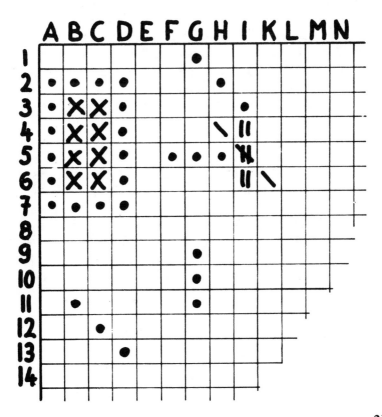

the location process has been completed. Any player, on his turn, can declare, for instance "The squares B, C-3 to C-6 are occupied by a ship" (a speedboat in this case). The player can do this even before he has begun the attempt to get a "fix" on a square.

If a "declaration" is correct, the opponent has to confirm it and then the boat is removed from the game. If it is wrong, the player, for punishment, has to declare the location of one of his own boats of equal size. If no boat of that size is available, he must declare the position of a smaller boat since it's more difficult to get a fix on the smaller boats. If no smaller boat is present, he must give the location of a larger one. Of course, this means he loses that ship.

What makes this version so intriguing is that the players must use their imagination and reason. For instance, in the illustration, squares I, K, 4, 5 must be part of a ship—either a U-boat or a larger ship extending to the right. How does one know? I-4 *to* I-6 is a bearing with two fixes—either I-4 and 6, I-5 and 6, or I-4 and 5. First, it is not possible for I-4 *and* I-6 to belong to two different ships. Since every boat occupies at least 2 squares, I-3 would then also have to be part of a ship, and this space has been declared to be "water." Further, I-5 and I-6 can be discounted, because such a ship would contain K-5 and K-6. In the bearing H-4 to K-6 only one "fix" was acknowledged. Since I-4 and I-6 cannot be fixes, I-5 must be one of the two in that bearing and thus the only fix in the intersecting bearing. Squares H-4 and K-6 as well as I-6 must be "water." Squares I, K, 4, 5 therefore have to be either a U-boat or part of a larger ship. Having established the situation in this manner, one need only determine the length of the ship involved to declare its location and eliminate it from the game.

BLINDFOLDED

Degree of difficulty: 3
Players: two

The object of this game is to find the treasure that your opponent has hidden *before* your opponent finds yours. This is a version of the pencil-and-paper game called "Sniff," invented by Mark Berger.

As with "Sink the Ship," each player has two pieces of graph paper, each with a field of squares 10 × 10. Identification is the same also: on the top, A to K, and on the side, 1 to 10. Make arrangements for a shield so that your opponent can't watch you as you bury your treasure.

Choose a square, anywhere in the field, and mark it with a circle. This is where your treasure is buried. Mark another square with a cross, the square where your opponent has to begin his search. To make the game more interesting, outline a maze like the one shown in the illustration on the following page. Both players should agree on the size of the maze, which in the beginning should not exceed 30 individual hurdles. A treasure hunt with 40 or even 50 hurdles is of course much more difficult. Whatever maze you choose, make sure that the road leading to the treasure is not obstructed.

Once all the preparations are completed, the players inform each other as to the location of the starting square, for instance D-2. Now the hunt can begin, with the players taking turns. The player whose turn it is will state in which direction he wants to move one step: up, down, left, or right. The opponent has to let the player know if anything is in the way. If the way is clear, he can proceed. A maximum of five steps is allowed per turn. However, the player loses his turn as soon as his progress is blocked by a wall of the maze or he has taken his five steps.

Every square can be touched as often as necessary, because when the hunter gets stuck in a dead end, the only way out is by backtracking. Keeping track of movements, therefore, is very important. This is particularly true for every movement of your opponent, in order to avoid giving wrong replies by

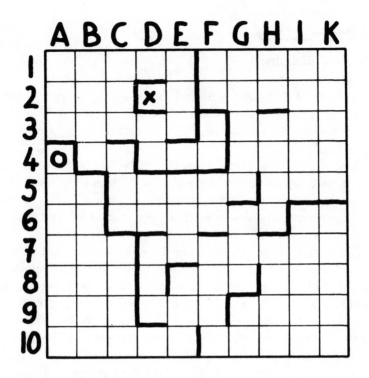

mistake. For better control, it is advisable to state the position and the direction of movement at the beginning of every turn—for instance, "I am at F-7 and will go to F-6." This assures both players that they have the same information and they both know where things stand. Also, possible errors can be corrected immediately.

ESCAPE

Degree of difficulty: 3
Players: two

Although players in most of these games start out as equals, in this exciting game of "Escape" the opponents are anything but. Here, one is the hunter and one is the hunted. And while the hunter is allowed to move in big leaps across the board in order to catch his victim, the hunted tries desperately to escape, which gets more and more difficult as the game progresses.

The drama plays itself out on a field of squares 8 × 8. This time, however, not the squares but the lines dividing the squares are identified as A to I and 1 to 9 (see illustration on the following page). Each player has his own field that he hides from his opponent's view.

After the players have decided who will be the hunter and who will be the hunted, the hunted takes position in the middle of the field, on point E-5. The hunter can hid in any one of the four corners: A-1, I-1, I-9, or A-9.

On his turn, the hunted can move either horizontally or vertically; however, backtracking to a point that was occupied immediately before is not allowed. Besides this, there are no other restrictions—meaning that the player can use every point in the field as often as desired.

The hunter, who can move freely from point to point, is also not allowed to backtrack immediately to a point he just occupied. Unlike the hunted, the hunter can also move diagonally, like the knight in a chess game.

The interesting aspect of this game is that the players move simultaneously, announcing their new positions to their opponents after every step. The hunter therefore only knows the position of the hunted after he has marked his first move. The hunter moves randomly in the hope that he will catch his victim, which of course is his objective. The hunted, on the other hand, is trying to get to safety, which he can find in two of the four corners of the field. There, he cannot be touched by the hunter, even if the hunter should enter that corner.

The game is played in two parts. In the first part, one player is the hunter and the other the hunted; in the second part, the players reverse roles. Each part can consist of one or more rounds, depending on how long it takes for the hunter to catch his victim. The player who catches his victim in the fewest rounds is the winner of the game.

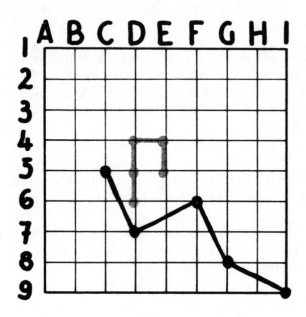

The first round is limited to 44 moves. If the victim is unable to reach safety by the forty-fourth move, the first part of the game is over and the players reverse roles. If the victim, however, was able to reach safety in 44 moves or less, the following rule applies: The number of moves needed to reach safety in that round minus one is now the number of moves available in the *next* round. The game thus becomes harder with each consecutive round.

Remember, this is not a game among equals: The hunter always has the advantage because he can move diagonally across two squares, while the hunted can only move vertically and horizontally the distance of one square.

GRID RUNNER

Degree of difficulty: 3
Players: two

The idea for this game came from a computer game. Two runners move along a grid. Every part of the grid they touch cannot be used again. This makes it much more difficult to stay in the race as the game progresses, particularly as the opponents try to upset each other.

The field is played on graph paper. The players each have a pencil in a different color. A field of squares 20 × 20 should be sufficient. The field is actually 21 × 21, since the runners are also able to make use of the field's outer lines.

The runners position themselves in the middle of the field, where the eleventh vertical and horizontal lines cross. One player marks the space of two squares in the vertical direction, one above and one below the eleventh horizontal line. The other player does likewise but in the horizontal direction. This creates a cross in the middle of the field, in one color on the horizontal axis, and in another on the vertical axis (see the illustration on page 28).

The race can now begin. Each player, on his turn, will move the distance of one square. With the first move, each player designates which end of the line is the head, because from now on the line can only proceed in that direction, with the head moving forward. The trail that the moving runner leaves

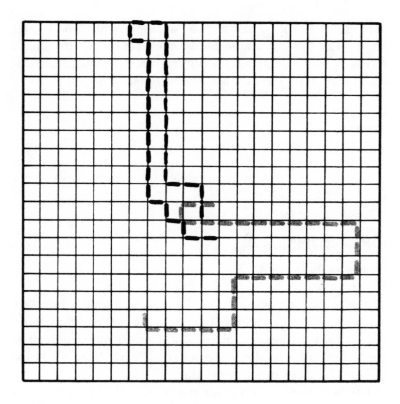

behind will look like a moving snake. The snake grows longer with every move the runner makes. A runner can go in any direction, but only along the grid lines, and only once on each individual line. A runner's line may touch that of the other runner (for instance, when a runner wants to make a turn), but the lines can never cross. The opponent's trail is like a closed border that cannot be crossed.

The object of the game is to stay in the field as long as possible, with each player trying to block off as much space as he possibly can. When a large area is blocked off, the runner can roam around in it, while the other runner has no way of getting into that part of the field. The player who runs out of space to move in has lost the game.

BORDER

Degree of difficulty: 5
Players: two

The game I am introducing here as a pencil-and-paper game has been on the market as a board game for many years under the name of "Twixt," and has made the inventor, Alexander Randolph, not only rich but also famous. A couple of years ago, a game called "Limes" appeared on the scene, but it was really no more than a rather cheap imitation of "Twixt."

The game, whether played as a board game or a writing game, is also known as "Border to Border"—referring to the sides of the playing field. The object of the game is to connect the two opposite borders. The size of the field is usually 10×10, but is sometimes expanded to 15×15.

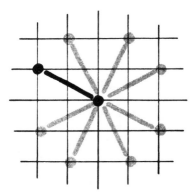

The game is played on the cross points of the squares. Both players select different colored pencils, which they will use to mark their progress during the course of the game. Each player "owns" two borders filling opposite sides of the board except for the corner squares. Each player marks his borders in his color (see the illustration on this page). The goal is to connect the two opposite sides with one continuous line while at the same time trying to prevent the opponent from doing so.

The players take turns, each using his own color. A move consists of coloring a free cross point with a dot (representing a tower). Whenever there are two or more dots of the same color exactly two squares diagonally apart from each other, a player can immediately connect them, providing that the line does not cross a line that is already present, regardless of its color.

The players are free to choose where they want to place their dots and lines; only the opponent's marked borders cannot be touched.

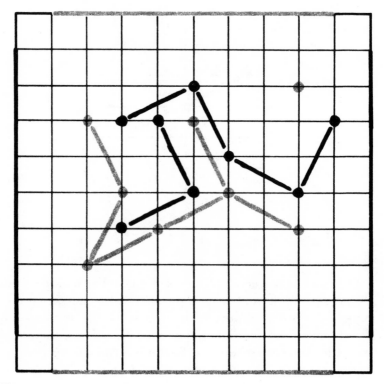

HONEYCOMB

Degree of difficulty: 2
Players: two

This is an easy game that can be brought to a conclusion in less than 15 moves. All you need are two pencils in different colors and a small piece of paper.

First, make six dots, of any color, on the paper. Now the players alternately connect two of the six dots with a line. Note that each line can only be drawn once, which accounts

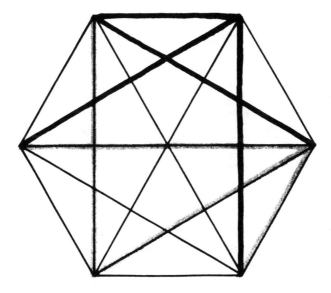

for the fact that the game is finished after the fifteenth turn. Furthermore, lines of the same color cannot form a triangle if all its end points are original outer dots. This does not apply to the small triangles that are formed within the honeycomb in the course of the game. The first player who is unable to connect two points with a line is the loser.

Should you choose to use seven points for the outer dots,

instead of the six-pointed honeycomb, again, only a small piece of paper is necessary, but three people now can play the game. In both versions, however, players should rotate at the beginning of each round, so that each time a different player

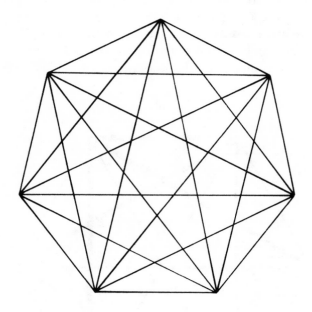

makes the first move. This is because when two people are playing with careful observation—and without making a mistake—the player starting a game can win every time, and when three people are playing, the first one is at a disadvantage.

GOMUKU

Degree of difficulty: 3
Players: two

"Gomuku" is a close relative of the game of "Go," which is usually played with "Go" figures on a "Go" board. "Gomuku" is an ideal game for learning the basic rules and tactics of "Go." However, since "Gomuku" doesn't require any figures, it makes a great pencil-and-paper game.

The game is played on graph paper. No boundaries are necessary, but if you do desire a border, a field consisting of squares 15 × 15 is the minimum. (If experienced players are

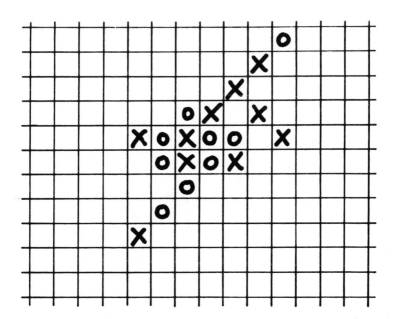

"squaring off" with each other, a field defined by a border can end in a draw.)

The players can either choose to play with different-colored

pencils, or one can use a cross while the other uses a circle. The players take turns marking any square that is still free. The object is for each player to mark five squares in a row; the squares can run vertically, horizontally, or diagonally. At the same time, the players try to prevent their opponent from doing the same thing.

The objective of this game has also given it the name "Five in a Row." The strategy of this rather simple game is to mark the squares in such a way so that they represent a multiple threat to the opponent. Note that a free square on either end of four consecutively marked squares already represents a win.

BURIAL

Degree of difficulty: 4
Players: two

The names for some of the most well-known games are often rather strange. This seems to be particularly true with games played by people congregating in a saloon or bar . . . . Why this rather interesting game of strategy is called "Burial" will probably remain forever the secret of the inventor.

With this game, the size of the field is unimportant, although it's a good idea in the beginning to start with a field of squares 6 × 6 or 7 × 7. Both players can use pencils of the same color, but different-colored pencils do make for a more interesting display on the paper.

Each player, on his turn, fills in one square. If that just-filled-in square completes an entire row—either horizontally, vertically, or diagonally—the player receives as many points as there are squares in that row. A row must consist of at least two squares and reach from one border of the field to another.

It is not unusual in the later phases of the game, for a player who fills in a square on his turn to close more than one row. The best a player can hope to complete is four rows: one

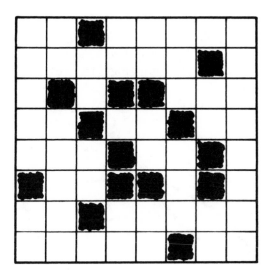

vertically, one horizontally, and two diagonally. In any case, the squares in each row are counted separately; the sum of all the filled-in squares in a row is the total count. The game ends when every square is filled in. The player with the highest number of points wins the game.

SPROUTS

Degree of difficulty: 5
Players: two

If people could be rated on a scale that would measure their enthusiasm for game playing, mathematicians would probably rank at the top. Maybe the reason for this is that mathematics itself can be viewed as a huge structure-creating game. Mathematicians often try to detect equations within and establish theories from well-known games, and many interesting games have been developed from the diverse areas in mathematics. One very fertile branch is topology, an area that deals with geometric structures.

From this area came the game called "Sprouts." Two mathematicians from the University of Cambridge, J. H. Conway

and M. S. Paterson, invented it on Feb. 21, 1976, during—how could it be otherwise?—their five-o'clock tea hour!

The rules of the game are fairly simple. All that's needed

are a large piece of paper and two pencils—the same color is okay. To start, randomly mark several dots on the piece of paper—the more dots, the longer the game. The number of moves necessary to complete the game depends on the number of dots placed on the paper. This can be calculated prior to the start of the game by applying the following equation: $m = 3d - 1$ (m = moves, d = dots). For instance, if the original number of dots was six, the total number of moves to finish the game would be 17.

The players take turns connecting any two dots, and then adding anywhere on that line a new dot. But there are a couple of restrictions. A line cannot cross itself or any other already existing line. And only three lines can converge on one dot, meaning that a new dot placed on a line can only accommodate one additional line. The winner is the last player who can connect dots with a "legal" line.

Since for each game the number of moves that conclude a game depends on the number of dots placed on the paper, the winner could be determined beforehand, providing that all calculated moves are being played. The strategy, therefore, is to find a way to reduce the number of moves that have been, or can be, calculated. That can be achieved by isolating existing dots by strategically placing the lines that connect them. It takes some practice to become familiar with this unusual strategy, but once you do, the game can become so fascinating that you won't want to stop.

FOG IN THE NIGHT

Degree of difficulty: 5
Players: two

During World War II, the owner of a lumber company in Holland invented a war game that he called "Stratego." However, it was said to be so complicated that many insisted it was unplayable. Some years later, "Stratego" appeared on the market in a less difficult version. The pencil-and-paper game "Fog in the Night" is based on the same idea.

The uniqueness of this game lies in the fact that while the location of the enemy is known, its fighting strength is not.

The battlefield of squares 7 × 7 is on the table between the two opponents. The columns have letters from A to G and the rows have numbers from 1 to 7 so that all 49 squares can now be easily identified. The five middle squares of the opposite base lines are numbered 1 through 5 and signify the starting positions of the five soldiers of each platoon. Three of those five squares are highlighted. These are the targets the enemy is aiming to reach.

Each player will keep track of the movements of all 10 soldiers on a piece of paper that he shields from the view of his opponent.

Draw two blocks and divide each into five columns—with two top rows, one for keeping track of your own troops, and the other for keeping track of your enemy's troops. Label the top rows from 1 through 5 to correspond to the numbers in the starting blocks (see the illustration).

Before the games begin, both players give each of his soldiers the value of his strength, and record these numbers in the columns of the right block.

A platoon consists of one spy and four soldiers. The spy has a value of zero, while the value of the other four soldiers must have a combined value of 12 and each must have a value of at least 1. For instance, you might choose to make number 3 the spy (with a value of 0) and give the first soldier a value of 2, the second soldier a value of 4, the fourth a value of 5, and the fifth a value of 1. The players keep their designations a secret.

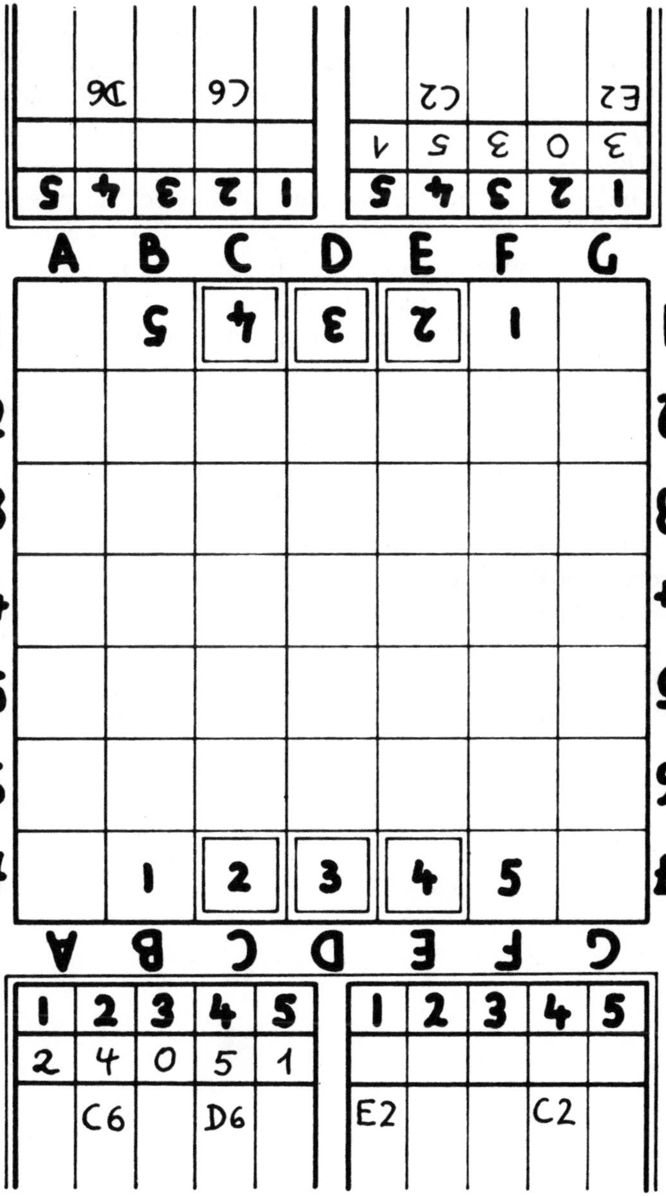

A B C D E F G
1 2 3 4 5 6 7
1 2 3 4 5
2 4 0 5 1
C6 D6
1 2 3 4 5
E2 C2

On each turn, the soldiers move one square, horizontally, vertically, or diagonally. Each position that a soldier is moved to is recorded in the respective column in the left block; for instance, "soldier 3 to F-4" might be recorded and announced to the opponent, who, likewise, would make a record of it in his right block.

Since moving only one square is allowed, no jumping is possible. Furthermore, each square can only accommodate one soldier. Two opponents meeting at the same block means confrontation. The platoon leaders now must give the strength/value of their respective soldiers. The soldier with the higher strength/value wins. If both soldiers have the same value, it is a draw and they must go back to their starting square. However, if this square has been conquered by the enemy, the soldier becomes a casualty.

The objective of the game is to either eliminate the spy, who is the figure with the lowest value and can be beaten by every one of the other four soldiers, or to successfully reach one of the three highlighted squares at the opponent's base line.

Experienced players, by the way, do not need the actual outline of the field; it is provided here for easier orientation.

More Strategies

BUILDING BLOCKS

Degree of difficulty: 1
Players: two or more

The game "Building Blocks" is, next to "Sink the Ship," *the* classic pencil-and-paper game. The mental capacity required for this game is just about zero, which makes it a favorite for

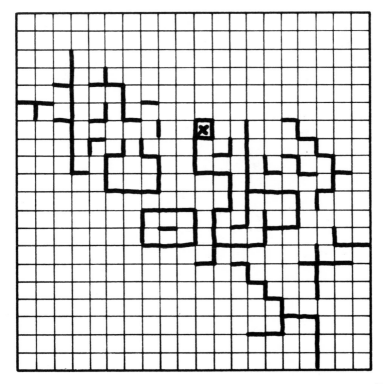

students to play under their desks during class. It is also an ideal game for passing the time on a rained-out vacation day.

The game is played on graph paper. Although it is best to use a large sheet of paper, the size can be determined by the preference of the players. It is customary to outline a rectangular field, but any other shape will do just as nicely. Each player has a pencil, of either a different color or the same color. Each player, on his turn, has nothing more to do than mark one side of a square. Whoever "closes" a square, on his turn, can either fill in the square with his color or mark the square with his initials; then he must make the next single line. The border of the field is regarded as one long line drawn before the game begins.

The player who has been able to close the greatest number of squares is the winner. However, during the course of the game, every player will try to position his line in such a way as to prevent the next player from closing a square. This way, a huge maze is created, and in the end, not unlike a domino effect, an opponent can close one square after the other.

If the game becomes too boring, try the "turbo-version." Here, the booty is not a single square but any conceivable shape that can be closed with one move.

SQUEEZE

Degree of difficulty: 2
Players: two or more

"Squeeze" is somewhat similar to "Building Blocks," because the object is to be constantly on the lookout for an open square.

The game, again, is played on graph paper, and the players can all use a pencil of the same color. Before you begin, outline a rectangular space for the "squeezing" to take place.

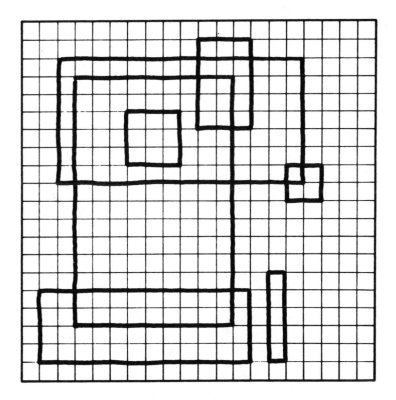

Each player, on his turn, has nothing more to do than outline a rectangle or square on the playing field. The size of each shape can vary, but the borders of any shape cannot be the border of the playing field, including its corners. Furthermore, every line can only be covered once, meaning that edges and corners are not allowed to share a line with another outlined shape. The individual rectangles or squares, however, may overlap.

In the beginning, it is easy to find space for outlining a shape. But after several rounds, players will begin to squeeze each other out of space. The player that is unable to find room and complete a shape is out of the game. If the next player, however, is more alert and can squeeze out another shape, he is declared the winner.

STAY WITH IT

Degree of difficulty: 3
Players: two or more

This little game comes from the magic bag of tricks of that preeminent author of books on games, Sid Sackson, who origi-

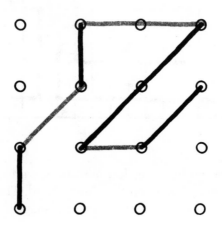

nally introduced it as a game for two players. But it can just as well be played with three or more people.

For two players, mark a piece of plain paper with 16 dots in the form of a 4 × 4 dot grid. Increase the number of dots according to how many additional people participate in the game. The color of the pencil used is not important.

The object is for every player, on his turn, to connect two or more dots with one straight line—either vertically, horizontally, or diagonally. However, the line cannot go around a dot.

The player who starts the game can place his line anywhere he wants. The next player must start his line from either end of that line. The player who is unable to draw another line has lost the game.

STAR FLIGHT

Degree of difficulty: 3
Players: two or more

Is there anybody who hasn't dreamed of flying to the stars? This little pencil-and-paper game that I put together for a newspaper a few years ago is like an imaginary journey, undertaken with nothing more than a piece of paper and a pencil.

The piece of paper becomes the universe where we hang our stars, randomly, without paying much attention to the distance between them. If your "universe" is rather small, don't hesitate to put your stars close together. Twenty-four stars will do just fine.

The players all fly in the same rocket. The first player is the astronaut who will circle the first star. The next player flies to the next nearest star and likewise makes a circle around it.

Traffic regulations in the galaxy are rather simple. First,

each player must proceed to the nearest star on his turn. Second, a route taken cannot be repeated on a player's next turn. Third, each star can only be visited three times. The player who is able to make the third circle around a star is rewarded with one point—and, of course, the star, which is now taboo. When every star has been circled three times, it's time to head home to Mother Earth.

The game ends when the spaceship cannot find another unclaimed star. The astronaut who has "collected" the greatest number of stars is the winner. It is not infrequent that arguments ensue over the question of which is the nearest star. Those who can be objective seldom have a problem with that question, but who can be objective when winning or losing a star is at stake?

There are two possible solutions to this dilemma. If democracy prevails, players simply vote. But usually those who are

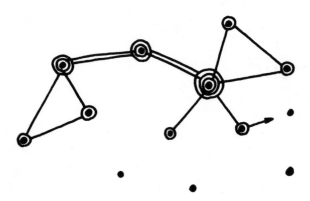

behind in the game will take sides against the player who is ahead. The other alternative is not quite as easy, because you have to measure, but geometry can be of help. A star (C) will not fulfill the necessary requirements if its position falls within an isosceles triangle whose base line is the distance of a "flight line" a player is contemplating to fly (from A to B) and whose height is one third of this distance.

In the diagram below, the distance between points A and B eliminates point C, since it falls within the isosceles triangle. Point D is okay, because its location is more than 30° removed from the imagined "flight line." Point E likewise is okay, because the distance from point A to point B is only slightly greater than the distance from point A to point E.

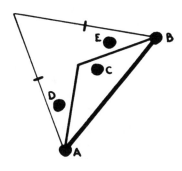

Considering these complicated astronomical measurements, it might be better to graciously compromise.

A WINDING ROAD

Degree of difficulty: 2
Players: two or more

This game was invented by William Black, which is the reason it's also called "Black's Road Game." In addition, however, it's also known as "Snake's Road" and "Squiggly Road." That's how things go in the world of games—very seldom is a game known by only one name, and it's even more rare that the author is known. Imagine the response of writers, let alone that of libraries and archives, if the same book had several different titles, and the author was either not known at all or his name just remembered by chance.

This game is played on graph paper. The field shouldn't be smaller than 5 × 5 but no larger than 8 × 8. A smaller field would be too confining and a larger one would only lengthen the game without adding any more suspense. This at least is true when two people play the game, because that's how Black

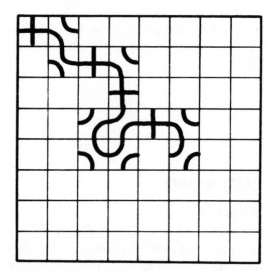

originally introduced it. Experience, however, has shown that this game can easily be played by three or four people. When this is the case, the size of the field should be increased. For four people, increase the size to at least 6 × 6, but no more than 10 × 10.

The players take turns marking a free square with one of three designs: a cross or one of two sets of opposing quarter-circles, as shown in the illustration. The first player places a design of his choice in one of the four corner squares. In subsequent moves, the players place one of the designs in an empty square. However, they must comply with the following rule: A design in a free square must be marked in such a way that the line began by the first player in the corner remains a continuous line until the end of the game.

Parts of a design disconnected from the continuous line can, during the course of the game, be incorporated into it.

With every turn a player completes, the road continues to grow until one of the players, on his turn, has the misfortune of "hitting" the edge of the field. That makes him the loser!

SLEEPY HEADS (RESTLESS NIGHT)

Degree of difficulty: 4
Players: two or more

It seems that people who construct pencil-and-paper games assume that these games are designed for only two people. So it is with this game, given to us by Alex Randolph, one of the few truly great inventors of games. Yet, "Sleepy Heads" can easily be played by three or more people.

The game is played on graph paper that is used as a dot grid. For a two-person game, the inventor suggested a grid of dots 6 × 6. Increase the number of dots according to the number of people participating.

The game is played with "sleepy heads." Each one consists of four connected lines that stretch over five dots. One of the two dots at the end is circled, to indicate the head.

During restless nights we toss and turn a lot. So it is with

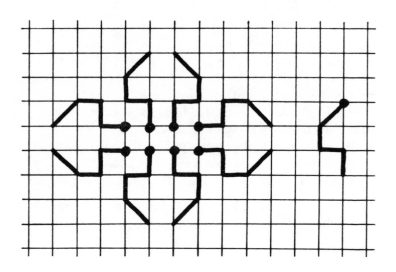

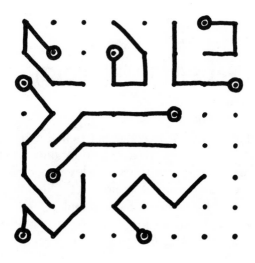

sleepy heads; they can contort themselves into 46 different positions during one of those restless nights.

In the game, each player, on his turn, has to position a sleeper on the grid. The following rules prevail: The heads of two sleepers cannot be opposite each other, either vertically or horizontally (maybe they all snore too much?). Furthermore, each of the 46 sleeping positions can only be used once, and decency requires that every dot in the grid can only be occupied once—meaning that sleepers cannot lie across each other but must be neatly folded into bed like sardines. Whoever is unable to squeeze another sleepy head into the bunk has lost the game. Good night!

L.A.P.

Degree of difficulty: 5
Players: two or more

The rather mysterious title of this game is actually formed by the initials of the man who invented it, the Polish author L. A. Pijanowski.

If the game, as intended by the author, is played by two people, it is a rather straightforward game of strategy. The investigation can also be undertaken by three or more people; however, when this is the case, a component of diplomacy is added to the proceedings and political savvy just might triumph over a sharp mind.

The secret proceedings take place on a field of squares 8 × 8. They are, as with other games in this book, labelled for identification with letters and numbers. Before the start of the game, each player prepares two 8 × 8 fields on graph paper in order to keep track of his opponents' as well as his own movements.

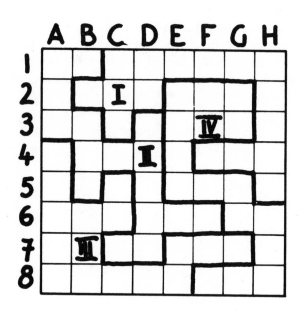

The field for each player's own use is divided into four irregular sectors labelled I through IV—each containing 16 squares (see the illustration).

The object of the game is to identify the opponent's secret sectors. On each turn, one player calls out a letter-number combination for four squares—for instance, D-E,2-3. The opponent's answer consists only of an acknowledgment if any portion of the dividing lines between sectors is involved; he will not, however, reveal the actual sections involved—for instance, sections II and IV. Each player has to determine the actual dividing lines between his opponent's sectors by repeated and overlapping questions and guesswork. Areas consisting of only one sector should be considered as the starting point of the investigation.

Players take turns. The player with the sharpest mind who has detected the enemy's secret sectors is the winner. Any attempts to name the squares belonging to a sector is allowed without any punitive consequences. The player who is being challenged should listen to the "guesses" of his opponents with a "poker face." When the guess is correct, he has only to answer, "Correct." Trying to name the squares can be a way of getting new information. However, since naming the squares belonging to a sector takes time, it shouldn't be done too often.

Several different ways of playing the game can be established when three or more people participate. For instance, it is possible for players to agree beforehand to take turns in calling out the suspected squares of a sector of another player, since a challenger is free to choose, during the course of the game, which one of his opponents he wants to ask. With such an arrangement, it is possible to gang up on one player, concentrating on an "investigation" until all sectors are identified and the player is eliminated from the game.

Another way of playing the game is to have the squares named by a challenger require a reply from every one of his opponents. This way, every player is able to gain valuable information about every opponent, and any player who has become a target has a better chance for defensive strategies.

TEMPLE

Degree of difficulty: 5
Players: two or more

The space for "Temple" should consist of squares 8 × 8. Increase the space when three or more players are involved.

First mark the space on a squared piece of paper. With the players taking turns, each chooses a different-colored pencil and marks a free square with a dot. Should there be a scarcity of colored pencils, assign a different mark, like a number, letter, cross, or circle, to each player. This arrangment, however, makes the proceedings somewhat cumbersome. It will already be difficult enough, as the game progresses, to keep track of the temple constructions.

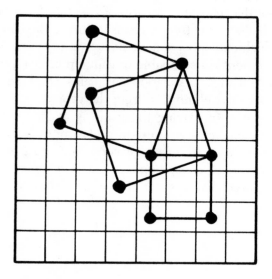

The squares that have been marked with a dot become the cornerstones of the temples. The player who is able, by connecting four dots, to complete a rectangle consisting of four grid squares has not only successfully built a temple but has also earned a point. The temple walls do not coincide with the grid of the graph paper. They can even run diagonally, as long as there is equal distance between the dots. Furthermore, one cornerstone can be used for many different temple constructions. If a player is able to complete several temple constructions with the positioning of one dot, points for each completed temple are awarded.

The size of the space on which the construction takes place should be such that each player has an equal amount of squares available on which to build. For three people, for instance, the "building site" should be 9×9. When the last square has been filled, the various builders mark and count the temples they have constructed. The size of the temple is not important. Each one counts one point. The greatest number of temples built by one builder determines the winner.

For a variation, the players can agree beforehand that as soon as a player has completed a temple, he is allowed to add another cornerstone. This will add excitement to the game.

HEPTAGON

Degree of difficulty: 3
Players: three

Before the game begins, take a piece of paper and draw an equilateral heptagon with a lead pencil. It's also a good idea to draw every possible connecting line beforehand. Try to be exact so that the pattern created by the connecting lines will give you a correct division of the space. The 21 necessary lines will yield exactly 50 sections.

The object of this game is for each player to claim as many sections as possible within the heptagon. The players, in the end, will identify their "spoils" with the color assigned to them. In addition to showing the players conquests, the colors create a wonderful picture. For that reason, it's nice to choose either complementary or contrasting colors.

In the game players take turns connecting two of the seven corners. Of course, each connecting line can only be drawn once. Since only 21 connecting lines exist, the game consists of exactly seven rounds.

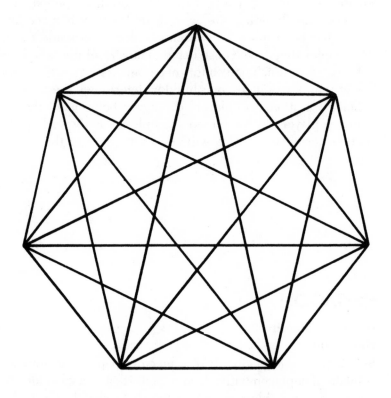

Evaluation of the outcome begins when every line has been marked. Each of the 50 sections is counted separately. The lines have created triangles, rectangles, pentagons, and in the middle, a heptagon. The player whose colors outline the greatest number of lines belonging to a section has won the

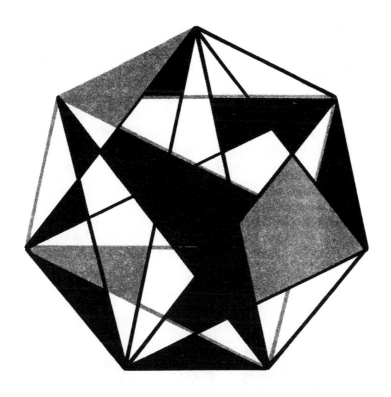

section and can fill it in in his color. In other words, a section
is awarded to a player who has marked more lines of that
particular section than any other player. However, if two
players have each marked two sides of a rectangle, or if three
players have each marked three sides of a triangle, the battle
over the space is considered a draw and nobody can claim that
section. The winner is the one who has filled in most of the
sections. Size or shape are not taken into consideration.

MULTICOLOR

Degree of difficulty: 3
Players: three or more

As already mentioned when discussing the game of "Sprouts," many pencil-and-paper games have their origin in the branch of mathematics called topology. One of the basic questions asked in this special field is how to deal with the "three-color problem." At issue is the attempt to prove the assumption that any area can be divided into separate sections with the use of no more than three colors without any two adjacent sections being the same color. "Multicolor" is based on this problem.

On a piece of paper, draw a field of squares 13 × 13. Players take turns coloring in rectangles within this field in their

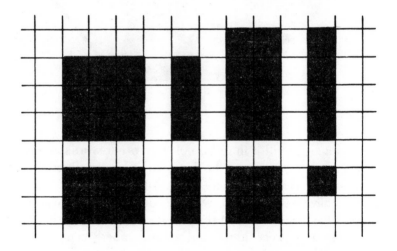

different colors. The rectangles can be as small as one square, but cannot be larger than nine. The area has to be chosen in such a way that its border does not include more than 12 sides of the squares that comprise it. See the illustration for some possible combinations.

Each player can choose the location and size of his rectangles among the still-open squares. However, the area claimed cannot come in contact with another area of the same color except at the corners. The player who is unable, on his turn, to claim a rectangle has lost. The end result of this game is a wonderfully colored field, assuming that the right color combinations have been chosen and the squares have been carefully filled in.

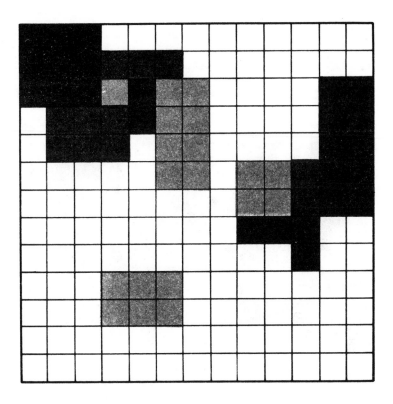

CRYSTALS

Degree of difficulty: 5
Players: two or more

"Crystals" is considered to be one of the most interesting pencil-and-paper games because it involves a combination of factors: strategy, imagination, beauty of form, and originality. Any number of people can participate, but—as is the case with almost every pencil-and-paper game—it is most fun when played by two or three people.

On a piece of paper draw a field consisting of squares 20 × 20. Each player is assigned a different color.

The object of the game is to build crystals that will occupy

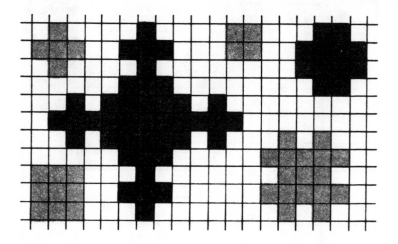

as large a space as possible. An individual square is called an "atom"; a collection of squares constitutes a crystal. The atoms must join on their sides, not just at their corners. The number of atoms to a crystal is not restricted, but each crystal must consist of at least four atoms. In addition, the shape of the crystal must be symmetrical—along the vertical, horizontal, and diagonal axes.

The crystal-building process takes place in the following

way. Each player, on his turn, puts an "X" in one square, or atom. A crystal can be claimed as such if its outer border includes at least four atoms and if one of them is of the same color. Crystals are allowed to touch other crystals—those be-

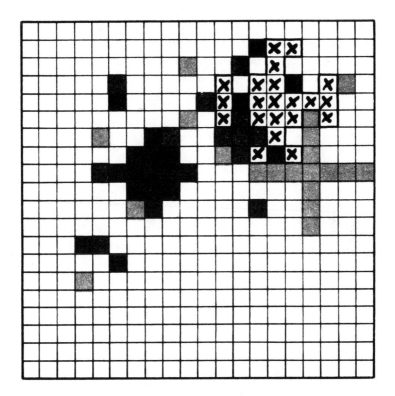

longing to an opponent as well as to the same player. The player who claims a crystal fills in all atoms in it with his own color (including those of the other players). Not only does this make for a pretty design but also for a better overall view of the field.

If all the participants agree that no more crystals can be built, meaning that nobody can hope to finish another crystal on the field, the game is over. Every player now counts the number of squares filled in his own color.

The rules for this geometrical game probably sound rather

simple; however, in the midst of all the commotion that ensues, much concentration is needed to detect the possibilities for crystal building. Also, it's not enough to try to color the greatest number of squares; you must always keep an eye on your opponent, and try to interfere with his strategy. Particularly in the beginning of the game, a lack of attentiveness can lead to an opponent being able to construct a crystal of considerable size. Later, when many atoms are already strewn over the field, and many crystals have already been established, you will be grateful if even the smallest of crystals can be built.

Action

FORMULA ONE RACING

Degree of difficulty: 4
Players: two or more

Auto racing has inspired many a board game, and pencil-and-paper games have not been overlooked. With multicolored pencils, a piece of graph paper, some talent for driving, and a good sense of how to maneuver curves, you can have as much fun with this game as at any actual race.

Amateur race car drivers should choose a piece of graph paper of considerable size for their race track. The actual race course should be several squares wide (see page 69); the shape itself is a matter of the players' own preference. Somewhere on the course mark the start and finish line. Just as on a real race track, the start/finish line should be in the middle of a long straightaway. The race is run on the grid of the graph paper. Players place their respective cars on the start/finish line. The race can be run in either one or more laps.

To begin, each car moves straight ahead the distance of one square, and this move determines the direction and speed of the next move. The length and speed of each move determines that of the next move, as follows: Extend the move by the same distance in the same direction to locate the center point that defines the options for the next move. This center point or any one of the eight surrounding it, can be the end point of that next movement. This scheme allows a player, on the next turn, to continue straight ahead or turn to the left or right and to put on the brakes or to increase the speed by choosing any of the nine points defined by the previous turn (see the diagrams for your moves starting from a dead stop).

The rules for the race are as follows. Players take alternate turns. The car positioned on the inside of the track makes the first move; the car positioned on the outside, the last. Cars are allowed to cross over the line of an opponent, but they cannot make a move that would define the same center point. If a player does this or makes a move so that the center point would be out of bounds he is forced off the track.

A driver whose car is forced off the track must, on his next turn, go back to the point that is closest to where he made the mistake. Center points that fall directly on the edge of the track are considered part of the track. The car that reaches the finish line first, after having completed all the agreed-to laps, is the winner.

For a "professional race," the rules are more strict. In following this version of the game, use graph paper with ⅛-inch squares. The race cars are positioned every 3 squares on the starting line. The cars basically move as described for the first version. But here the center point indicated by each move

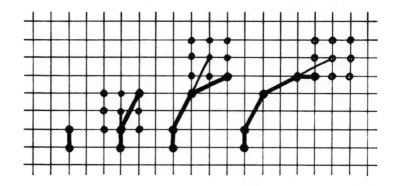

defines an area of 10 × 10 squares. The driver is allowed to choose any one of the 121 points within this area as the end point. As before, at the beginning of the race, the speed is zero, so a car on its first move is allowed only to move one square.

To show off his superior driving skills each driver has a certain number of "movement points," which have been

Start
Finish

agreed to before the start of the race. The number of "movement points" could be as much as 30 for one round of a race course of average difficulty.

Every player can use any of these points during a turn to increase the field within which a car can be manipulated. If a driver, for instance, uses 6 movement points, the field defined by the center point is not 10 × 10 squares but 16 × 16. Only an even number can be chosen on any turn. Using "movement points" allows the skilled driver in a strategic situation to either speed up or slow down his progress and increases the tactical possibilities the driver has at his disposal.

A space of three squares must always exist between any two cars and a car cannot drive across a position so occupied by another race car, or have as its end point such a point. A car in motion must be able to either speed up, slow down, or steer around another car.

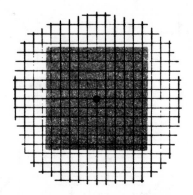

If a car speeds out of control beyond the track, the driver must, on his next turn, with painstakingly slow movements, try to get back on the track—meaning that he can only move three points at a time. When the track has been reached, the car must resume a speed of zero, the same speed it started with at the starting line.

Real pros add a further touch of reality to the game by

letting the cars move simultaneously. Depending on the number of race cars on the track, some very ticklish moments can be created.

In this version, every player has a piece of paper on which the anticipated moves of his car are recorded like coordinates on a graph. The papers are exposed simultaneously; then the movements are executed and analyzed.

The center point that defines a field of movement is like the center in a graph and the designated end points can be easily identified by their respective coordinates. Direction (the X-coordinate) is written down first. If it is a positive number, the movement is to the right; if it is a negative, the movement is to the left; if it's "0", the movement is straight ahead. The speed (the Y-coordinate) is written down second. The number is positive when the end point within the field of movement is farther ahead; the number is negative when it is farther back; and it is "0" if there is no change in speed.

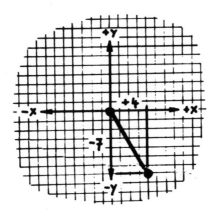

For example, a driver makes the following notes: "4, −7". The end point of his movement can be found by going from his center point, the point after the first phase of the movement, 4 squares to the right (+4) and then 7 down (−7). The car will move a little to the right and brake sharply. Since the normal size of a field of movement is only 10 × 10, the driver in this example must use 4 "movement points" to increase his field to 14 × 14.

When all cars move simultaneously, collisions are not uncommon. Should two cars end up at the same end point, a collision has occurred and the cars are so damaged that for them the race is over. If the distance between two cars is less than three squares, the collision is not serious, but the cars have to begin the next round at a speed of zero.

AUTO RACING

Degree of difficulty: 2
Players: three or more

In formula 1 racing much of the action plays itself out in the imagination. In "Auto Racing," the actual movement of the hand across the paper decides the winner and loser. Mathias Mala introduced this action-packaged game.

The fun lies in the movement of the hand and wrist. Get a feeling for this by following a loop that has been drawn on a piece of paper with a pencil. Repeat this over and over, constantly increasing the tempo, until the hand itself has absorbed the rhythm of the movement.

For this race, the track is laid out beforehand on a piece of paper, which does not need to be very big. The track should not be too wide and should ideally include a narrow pass. Curves can be established any way players see fit. The total track should only be large enough so that it can be traced with one hand.

Measuring and evaluating each run is the most difficult task

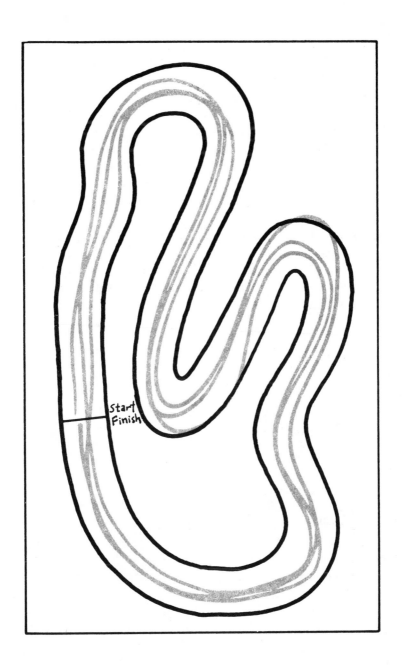

Start
Finish

of this game. But maybe that is really not all that important. Several different possibilities are available for competing.

If carbon paper or a copier is available, several identical tracks can be created and every player can have his own track. Each player positions his pencil at the starting gate. It is advisable to have a race track official who can give the starting sign and who can keep an eye on the participants. On the word "go," cars fly over the track. If a car driver runs off the track, the car is out of the race. The driver that has made no mistake and finishes first in completing the required number of laps is the champion.

If a stopwatch is available, the game can also be played on one track that every competitor has to maneuver in turn. Rather than disqualifying a driver who has run off the track, deduct a second from the total time for every square that the car strays from the track.

MANURE HEAPS

Degree of difficulty: 3
Players: two

This is a game that requires little strategy but is lots of fun. Two players try to circle several manure heaps with a wheelbarrow overloaded with—manure—what else!

First put down 14, 16, or even 20 consecutive numbers on a fairly large piece of paper. Both players should use the same pen or pencil so that one will not have any advantage over the other. A thick felt-tip pen is ideal. The pen is not held in the conventional manner but rather by the very end between the tips of the thumb and the index and middle fingers. After all, we are dealing with manure and who wants to get dirty fingers!

Decide beforehand who is to start this smelly race. One

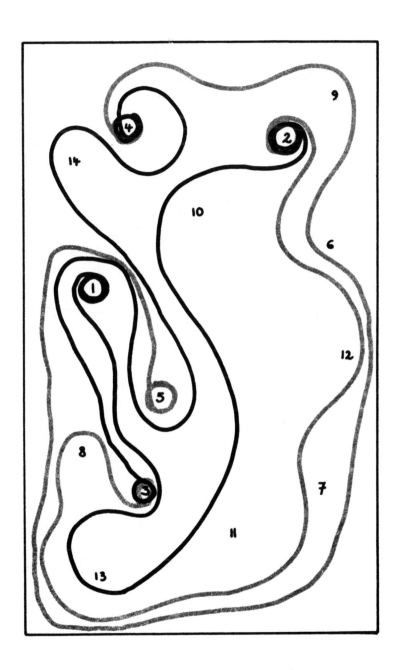

player circles number "1" and proceeds in a random movement over the paper, finally circling number "2", gaining a point. The pen is now passed to the next player, who circles number "2" and, like his opponent, rides in big loops over the paper to number "3" and circles it. This continues until the last number has been circled.

The trick to this country race is that, for every mistake, the player gets one point deducted from his score. It is absolutely forbidden to touch an already existing line or to cross over one. And this is not all that easy, considering that you are holding the pen at the very end with your fingertips.

The numbers are the manure heaps, giving this game its funny name. By the way, players are allowed to drive through these manure heaps anytime without getting minus points—hold your nose.

PIRATES

Degree of difficulty: 3
Players: two or more

There is a great deal of difference between a tax collector who confiscates a shipload of goods and a lawless pirate who attacks and plunders an innocent ship. Or, is there? Whatever the case may be, men, raise the black flag with the skull in the center!

As always, the field of action—here, one of the five oceans of the world—is laid out on a piece of paper (see page 77). The size of the paper is important, particularly if more than two people play the game, so be prepared with a large piece of paper for the upcoming battle.

First, each pirate draws an island in the ocean, with a safe

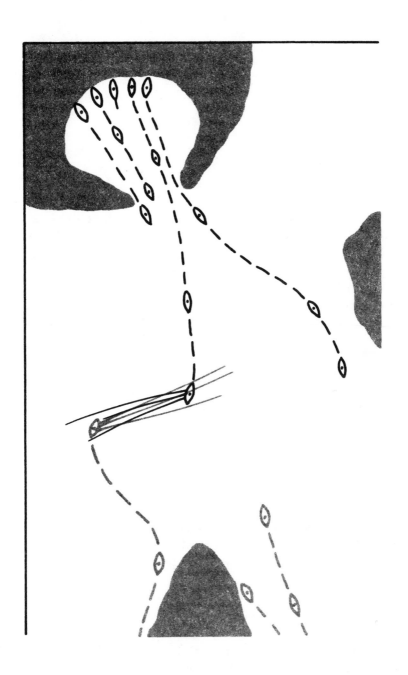

harbor for his fleet. The distances between each harbor should be about equal and each should be able to accommodate five ships. The individual boats are almond-shaped with a dot in the middle. One of the ships in each fleet is outfitted with a galleon mast.

There are several different versions of this game. A rather simple version has as its only goal defeating the opponent by sinking the flagship, the one with the galleon mast.

In a more sophisticated version, opponents number their ships and assign each a secret value. The combined value of a player's five ships will be 21. Whereas in the first version players are eliminated from the game when their flagship has been lost, here the object is for each participant to reach the opponent's harbor, or any other previously designated point. When a ship has done so, the pirate is awarded the points he previously assigned to that particular ship. The ship can still move about the ocean, but its points can only be collected once.

The game is as rough as the sea. On each turn, as many ships as desired can be moved about but no farther than for a total of seven units. When a ship sets sail, its progress is marked on the paper with a broken line, with each dash measuring about ⅜ of an inch. The final unit of the move is the ship itself. Again, the maximum length of a turn is seven units. Any pirate, however, can choose to take only a portion of these units or not to move his ships at all.

At the end of each turn, the captain is free to fire on his opponent. To do this, he holds the pencil at the very end with the tips of his fingers. After placing the point on the side of his ship, he flicks the point towards the target with a flick of the wrist. If the shot lands on the center of the target ship, it has been sunk. But if the target survives the attack, immediate revenge is taken. The attacker cannot shoot again, until the attacked ship has had a chance to shoot back.

If a ship has been hit without being sunk, it cannot be moved during the pirate's next turn. In other words, after an exchange of "fire," a ship that has been hit must "sit out" one turn. If a ship has not been hit, the owner may, in the next round, move close to the wounded ship and successfully send it to the bottom of the sea with one shot.

Shots can only be fired across the open sea. Should a shot cross land, it automatically falls into the water. Since the players in the beginning agree on how many cannonballs each player has in his arsenal, an accurate account of the number of shots fired must be kept. One hundred apiece is a good number.

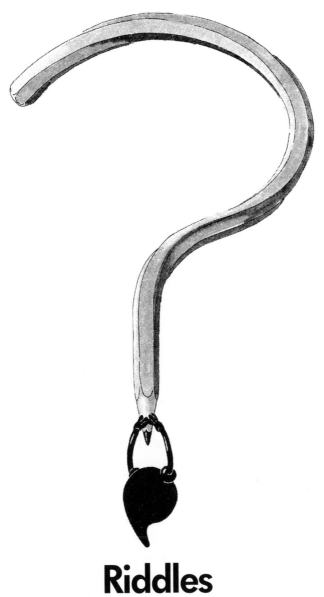

Riddles

THE SALESMAN

Degree of difficulty: 6
Players: one

In this game, the salesperson looks at a map where several locations have been highlighted. His task is to find the shortest possible route by which every location can be covered.

Even mathematicians have yet to find a sure-fire method to

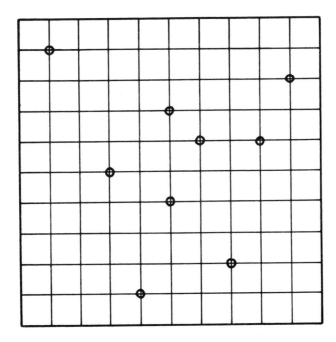

figure which route will be the shortest. That should give pause for thought. . . .

The game is played on graph paper; the size of the area is 10 × 10 squares. Nine circles are placed on intersecting lines on the graph. These are the towns that have to be covered on the trip. The route runs along the lines of the graph and cannot go diagonally through a square. Furthermore, each line belonging to one square can only be covered once on the trip to all nine towns. The driving route can, however, be crossed whenever necessary. The end and beginning of the trip can be anywhere and the end of the trip does not have to be at the starting point. The object is to find the fastest route by which the driver can cover all the towns.

DROODLES

Degree of difficulty: 5
Players: One or more

A droodle has been defined as a "picture riddle," or, more accurately as a "doodle riddle," hence a "droodle." Whether the word is a deliberate coinage or has legitimate etymological roots, "droodling" has been around a long time.

Anybody who knows "droodling" understands its singularity. Although a good "droodle," like a good riddle or a good joke, is almost impossible to explain, each usually has something satiric, whimsical, or tragicomic at the core. For anybody unfamiliar with droodles, the best way to understand them is to try and solve the ones on the opposite page. A look at the answers on page 86 will make everything clear.

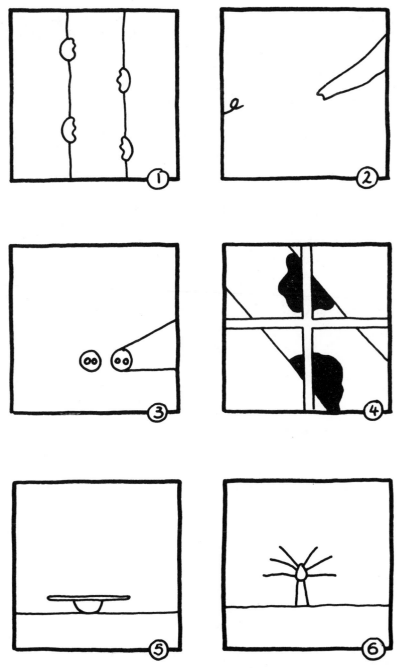

Now try your own hand at some "droodles." There are really no formal rules beyond the tradition that "droodles" usually begin with a square frame and that the drawing should be as spare and simple as possible. The art form is simple enough, however, that even these simple rules can be broken if that is necessary to produce a really dry, corny riddle or brilliant, biting gag.

Competition is like any riddling contest and can be played anywhere by any number of contestants. Once you present your opponent with a "droodle," he can try a solution orally or write it on a piece of paper. The "droodler" should answer only "yes" or "no" and not divulge the solution until the whole contest is over and all opponents are stumped. It can sometimes prevent controversy if the solution is written on the back before the game begins.

In a game with several players, a player takes possession of any "droodle" he has solved and the player with the most "droodles" at the end is the winner.

Answers: 1. A bear climbing the other side of the tree. 2. An elephant chasing a pig. 3. A pig looking for an outlet for his impulses (electrical). 4. A giraffe passing a window. 5. A sunny-side-up egg flat on its back. 6. A spider doing a handstand.

ANAGRAM

Degree of difficulty: 5
Players: one or more

The dictionary defines "anagram" as "a word or phrase made by transposing the letters of another word or phrase" and as "a game in which words are formed by rearranging the letters of other words or by arranging letters taken (as from a stock of cards or blocks) at random." To play, take a strip of paper and write a word, a name, or a sentence in big letters. Cut up the paper so that each piece shows one letter, and then rearrange these letters to make up a new word (or words).

A favorite version of "Anagram" is the game called "Calling Card." Take the words on a calling card—usually a person's name, together with secondary information such as an address or profession—and rearrange the letters into words describing either a hobby or a saying.

MAZE

Degree of difficulty: 4
Players: one or more

For Theseus, the maze of the Minotaur was a deadly game. Greek mythology tells us that it was Ariadne who gave him the thread that enabled him to win the game. And that mythical thread has since become world-famous.

Today, thanks to some very smart, observant people, we know that one can navigate a maze successfully if one keeps continuous contact with either the right or left wall of the maze. To cover a maze in the shortest possible distance, the hand-on-the-wall method is no help. There is another sure-fire trick that *will* help you systematically find the shortest route through the maze.

Look for the end of a dead-end section on the map. With a pencil, close off the road at the first intersection. Don't be fooled by turns. If only two roads come together instead of three or four, it isn't really an intersection.

Finding a dead-end section and then closing off the "road" at the first intersection is repeated until there are no more dead-end sections left. This means that only the very most direct route between the start and finish line is left open. Picture the maze as a treelike structure, where the cutting begins at the tip of the stems, and from there proceeds to the little twigs, and then to the branches, until finally the trunk is reached.

Inventing and solving a maze has been done by computers for a long time now. Luckily, however, it still can be done with pencil and paper.

The game is played on graph paper, with the walls of the maze being the lines of the graph; the size is determined by the maze builder. The road through the maze is the space between the lines and should always be the width of one square.

To make the maze, start by outlining a wall that will be the outer dimension. Next, choose the entrance and exit of the

maze. Both should be one-square wide. If the design is circular, the exit can be at the center of the maze instead.

Parallel to the outer wall of the maze, locate a point from which a second wall is drawn vertically. This wall can be drawn with any kind of turns and twists, and can be of any length the builder chooses. The wall, of course, cannot touch at any point—not itself nor any other wall. When the maze is finished, the walk through it must always be one square.

After completion of this wall, start anywhere at a point on the already existing wall and draw another wall according to the rules above. Without violating any rules, continue until all the possible squares have been incorporated into the maze. Now the maze is finished.

It is great fun to build a maze with a group of players. Colored pencils aren't necessary but will add to the excitement of the construction. When the maze is finished, use different-colored pens or pencils and fill in the dead-end roads, until in the end a white road will show the way out of the "a-mazing" picture.

Letters

TOWN, COUNTRY, RIVER, AND SO ON

Degree of difficulty: 3
Players: one or more

Each player has a piece of paper, marked with several columns. Players have to agree beforehand on how many columns and what categories they want to use: towns, countries, rivers, actors, politicians, pop stars, painters, composers, food, drinks, first names, building materials, tools, cars, hobbies, sports, plants, animals, chemical elements, medical prescriptions, office supplies; you are only limited by your imagination, and the number of columns is limited only by the size of the paper.

The players hide their paper from the view of their opponent(s). One player silently recites the alphabet until another player says, "Stop." The letter that the player has arrived at at that moment is announced. This letter is now the first letter of the names to be written under the category at the head of each column. For instance, if the letter "A" is announced, you would write Australia, Argentina, and so on, under the "countries" category.

The first player to finish every column with the appropriate words gives a signal that indicates the end of the round. The players now uncover their paper. Empty columns are marked with a dash. Every word that has been used by more than one player is crossed out. For every remaining word, the players get one point. It is not enough to quickly fill in the column; the words should be as unusual as possible.

The number of rounds to be played is left up to the participants. But the winner should be declared after no more than

24 rounds, because for every round a new letter is chosen, and "X" or "Y" should not be inflicted on any player. Usually the game ends much sooner, either out of boredom or because the paper is full. Therefore, it's a good idea to agree beforehand on the number of rounds to be played.

If no partner is at hand, this game can also be played by one person. Prepare a piece of paper as mentioned before and think up some interesting categories. On the left side write down the alphabet, from top to bottom. Now fill in the columns. For those who like a challenge, use a stopwatch to determine your time for each round.

INVENTORY

Degree of difficulty: 3
Players: two or more

The "Town, Country, River, and So On" game has many cousins. Let me introduce one of them here.

The players agree beforehand on one category—for instance, towns, first names, or flowers—and everybody tries to quickly find an appropriate word for every letter of the alphabet. Leave out the letters "X" and "Y" and possibly also "Z".

The first to finish the inventory signals the end of the round. As with the previous game, words used by more than one player are eliminated. For every other word, the players get one point.

FILL THE VOID

Degree of difficulty: 3
Players: two or more

As with the games "Building Blocks" and "Sink the Ship," "Fill the Void" is like a hardy evergreen: It never goes away. Here are the rules. On a piece of paper, write a word—on the left side from top to bottom, and on the right side from bottom to top, leaving room between them (see the illustration). This is the void that is to be filled with words that start and end with the letters already there. The object is to be the first to finish the task. For every missing word, the loser has to give the winner a point or a chip.

CASCADE

Degree of difficulty: 3
Players: one or more

It is possible to have a competitive race with words. It doesn't matter whether just one person plays the game until he runs out of steam or whether a whole group competes, either against the clock or according to a predetermined number of words.

In this cascade of words, the players all start out with the same word, which has been selected by the group. The word should not be too long; four or five letters are ideal (for instance, "hand"). Everyone writes this word at the top of his paper, and the cascade can take off. A word is written under the initial word, in which one of the letters is changed. The order of letters can also be changed. This is repeated line after line, always changing one letter and possibly rearranging the rest. The words will cascade until you've exhausted all possibilities. Of course, every word can only be used once (for instance, hand, wand, want . . .).

This is an ideal game to be played by one person. When played by a group, a time limit is set and the fastest "cascade" is determined, or a limit of 30, 40, or 50 rounds are determined beforehand and the winner is the player who finishes first.

SNAIL'S HOUSE

Degree of difficulty: 3
Players: two or more

With this game, the players all work on one word. First draw a snail's house that is divided into approximately 20 sections.

The number of sections should be divisible by the number of participants; so, for three players, for instance, there would be 21 sections.

One player begins the game by writing a letter in the first section. The paper with the snail's house is passed on from player to player, with everyone adding a letter in the next free section to the ever-growing word into the next section. The

end result must be a real word. If a player runs out of ideas and passes, the player who has added the last letter must reveal what the word was that he had in mind when he added the letter. This word cannot be longer than the sections in the snail's house.

The player who passes is the loser and has to pay for a round of ice cream.

SHAKING A WORD

Degree of difficulty: 3
Players: one or more

Most games have a certain time limit that determines the end of a round or the end of the game, and that establishes the winner. Games without a definite end are rare, but "Shaking a Word" is one of them.

Think of a word or a phrase that isn't too short—"writing game," for example. Every letter in these words, or just some of them, are to be used to create new words. The object is to find as many different words as possible: "write, writ, grit, gang, mat, mate, tame, grim, rig . . ."

When is the game finished? Probably not until the participants are exhausted. Whenever the final counting takes place, see how many different letter combinations can be constructed and/or how many actual words have been formed. If you include proper names and words from a foreign language, a small lexicon could be filled—or even a large one.

CROSSWORD PUZZLE

Degree of difficulty: 6
Players: one or more

This game has really very little in common with the well-known and much-loved puzzle in the daily papers.

Here, each player needs a piece of graph paper and a pencil. A field of squares 6 × 6 is marked off. When five to seven people participate, each player chooses one square, at any location. If more than seven people are playing, the size of the field should be increased by as many squares as are divisible by the number of participants.

Of course, at the beginning of the game, each player hides his paper from view. The rules are very simple. The player, on his turn, announces a letter that every player, including the one who announced it, will write into an empty space on his field. Of course, it is not enough just to set the letter blindly in a square. The object is to create as many real words as possible, vertically, horizontally, and possibly diagonally.

As is the case with every game, participants have to agree beforehand as to which words will be acceptable. If proper names are used, for instance, a clever player could insist that about any combination of letters could constitute a proper name. The easiest way to get around some disputes is to have

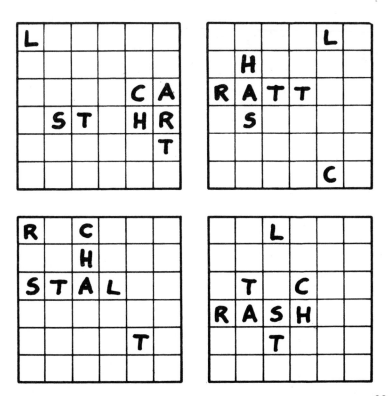

a dictionary on hand. There is no reason why you also couldn't have a foreign language dictionary at your disposal.

When every square in the field is filled, the players count the words that they have created. Every word counts as many points as it has letters. A word must occupy either vertically or horizontally connected squares. An individual letter in a square can, of course, be part of many different words. If very ambitious people get together, words created in a diagonal direction can also be included.

A person with lots of time on his hands, can play this game by himself. The object of this version is to create in a field of 6×6 squares words with the highest possible letter count.

Discounting diagonally positioned words and including two-letter combinations as a word, it is possible in theory to get as many as 600 points. The various squares in a 6×6 field can also have different value points. With this version, the players keep an eye on the designated values of the various squares when positioning words, horizontally, vertically, and diagonally.

Numbers

EVA MIDGET

Degree of difficulty: 4
Players: three

Eva Midget, the girlfriend of Adam Giant, the famous mathematician, is supervising three players. Each one has been given a piece of paper and a pencil.

The game begins with every player, in turn, announcing a number, so that they have constructed a three-digit number, which they then write down on a piece of paper. Randomly using the four basic mathematical operations of addition, subtraction, division, and multiplication, and by using any number from 0 to 9, they are to construct a mathematical calculation with the original three-digit number as the outcome. Within the calculation, every one of the 10 numbers can only be used once and can be used in combinations of no more than three digits.

The player that arrives at the originally established three-digit number is the winner. If nobody is able to reach this "goal," the player who, after a preestablished time, comes closest to the predicted solution is the winner. The winner in each round will be rewarded one letter from the name of Eva Midget. The player who has "collected" all the letters necessary for the name Eva Midget is the Grand Champion.

PAPER BOXING

Degree of difficulty: 4
Players: two

The title fight for "Numbers" is the brainchild of Sid Sackson, an American inventor of games.

The fight takes place in a ring of squares 4 × 4. Each player marks his own ring in secret. An "S" in the upper left square is the starting square; in the rest of the squares, the players randomly write numbers from "1" to "15". Next, players agree on who will fight with the even numbers and who will use the uneven ones. The numbers that occupy the lower left squares in the players' individual rings are added. If the number is uneven, the uneven fighter begins the game, and vice versa.

Starting from the "S" square, the first player draws a vertical, horizontal, or diagonal line to an adjacent square in his ring. The opponent now does likewise on his turn. Whoever

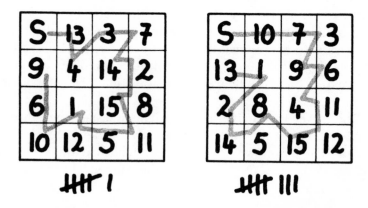

connected to the higher number has won the round and is awarded one point. If both numbers are the same, the round is a tie and the same player draws a line again to start the next round. The player that connects to the higher number starts the next round.

Only one step to an adjacent square that is still free is

104

allowed in each round. The line, therefore, can go in any direction but cannot end up twice in the same square. If a player maneuvers himself into a square from which it is impossible to reach another free square, the opponent has won the fight by a knockout, even with fewer total points.

NUMBERS JUMPING

Degree of difficulty: 3
Players: two

Don't be alarmed. This game doesn't call for mathematical calculation. Here, both players mark their steps on the same field. Jumping continues until one of the two players gets stuck.

The jumping takes place on a field of squares 5×5 (see the illustration on page 106). After having agreed on who will start the game, the first player marks a "1" into a square of his choice.

The opponent then marks any one of the squares in the same row horizontally or vertically with a "2", as long as the horizontal or vertical movement from the "1" to the "2" does not cross an already occupied square. In the beginning, of course, there's no problem because the field is still relatively empty. That, however, changes fast, because every square can only be occupied by one number.

Players take turns and always use the next consecutive number. One player, therefore, will always use an even number, and the other, an uneven number. Eventually, so many squares will be filled that a free square cannot be reached, either horizontally or vertically, without crossing over another number. The player who finds himself in such a predicament has lost the game. Since the whole field consists of 25 squares, the game can't last any longer than 25 turns.

			1	

3		4	2	
			1	
		5		

3	10	4	2	7
	9			8
			1	
		5		6

3	10	4	2	7
12	11			
	9			8
			1	
13	14	5		6

3	10	4	2	7
12	11	17		18
	9			8
	15	16	1	
13	14	5		6

3	10	4	2	7
12	11	17	19	18
	9	21	20	8
	15	16	1	
13	14	5		6

666

Degree of difficulty: 5
Players: one or more

Each player has a piece of paper and a pencil. They write randomly 13 numbers on the paper, from 0 to 9, with no number appearing more than three times. On a signal, each player passes the piece of paper in front of him to the player next to him. Each player now has 13 numbers in front of him that were chosen by somebody else. Combine these numbers into new numbers that are not to exceed three digits. Take these newly created numbers and form an equation within which any of the four basic operations can be used: addition, subtraction, division, and multiplication. Multiplication with "0" is not allowed, but the use of brackets is. The end result must always be the number 666!

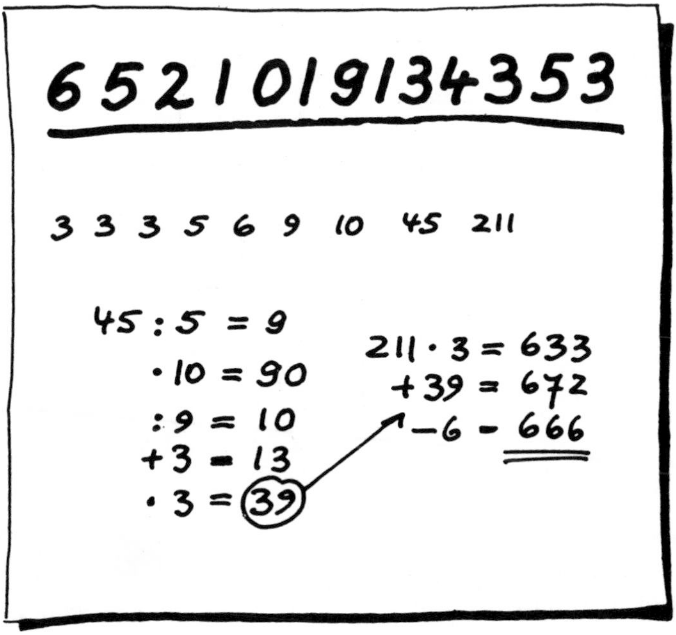

6 5 2 1 0 1 9 1 3 4 3 5 3

3 3 3 5 6 9 10 45 211

45 : 5 = 9
  · 10 = 90
  : 9 = 10
  + 3 = 13
  · 3 = ⟨39⟩

211 · 3 = 633
  + 39 = 672
  − 6 − 666

A DUEL WITH 12

Degree of difficulty: 4
Players: two

The configuration of the dueling field is difficult to explain without a diagram (see below). The number "6" is marked in the middle of this 25-square field. The other 24 squares remain empty. Each player makes a list of numbers from "1" to "12", which are used, one by one, and then crossed off the list.

The first player chooses one of the twelve numbers, crosses that number off his list, and places it in one of the empty squares in the field. When a player is able, by filling one square, to complete one of four 6-square value formations (see the illustrations), he is awarded a point. If a player doesn't complete a formation, the play passes to his opponent.

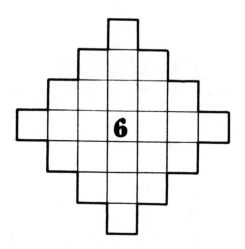

Each of the four 6-square formations has a different value: A row of six squares is worth "4"; a cross, "3"; and a rectangle or a triangle, "1". The four formations can be located in any direction on the field. If a player completes more than one formation on his turn, he can choose which 6-square formation he wants to have counted.

When a player completes a formation, he takes the sum of the numbers contained in that formation and subtracts that sum from the number "39". He then multiplies this difference by the value of the formation and this is the number of points that he is awarded.

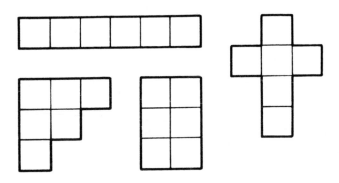

For those who think that these rules are not complicated enough, a restriction can be built into the game that allows each player to use a formation value only once.

The game is over when the last available square has been filled, and the player with the highest number of points is the winner.

THE SUM OF A SUM

Degree of difficulty: 5
Players: two

In this numbers game, both players start out under identical conditions and are equipped with the same tools.

Between them is a field of squares 4 × 4. Each player takes a turn writing any number between "1" and "8" into one of the free squares. Each number can only be used once by each player. To keep good control, a number that has been used is crossed off a list that has been prepared prior to the game. The field is filled when every player has placed all of his eight numbers on the field.

Now, one player adds the four numbers in the four vertical columns, while the other adds the four numbers in the four horizontal rows. The sum of these additions, however, is not what counts. What *does* count is the difference between each sum and the number 18. The amount of that difference is the player's total score.

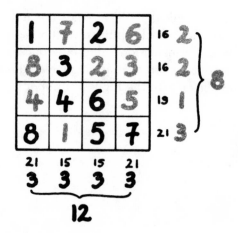

LOTTO

Degree of difficulty: 1
Players: three or more

This game requires an umpire who doesn't participate in the proceedings. That's why at least three people are needed to play the game, but the more the merrier!

The umpire needs 48 pieces of paper, on which the numbers "1" through "12" are written, with each number being used four times. The umpire puts all the pieces of paper in a container, draws out one, and announces the number that is written on it.

Each of the participants has a piece of graph paper on which he marks out a field of squares 5 × 5. The players use the squares to write in the number that the umpire announces. The object is to write as many like numbers in the same row or column as possible. But it isn't all that simple to

7	4	2	1	5	0
7	12	2	9	5	0
10	11	11	9	12	10
7	4	6	3	5	25
10	3	6	9	5	0
40	10	20	30	50	185

keep order in the squares, because only 25 of the 48 numbers are drawn from the container.

At the end of the game, when every square has been filled, the players add up their numbers to find their score. Every column and every row is treated separately.

A pair of like numbers and three different numbers in one

111

row or column earn 10 points; two pairs of like numbers and one different number earn 20 points; three like numbers and two different numbers earn 30 points; three like numbers and one pair of like numbers earn 40 points; four like numbers and one different number earn 50 points; five different numbers earn no points; and five consecutive numbers, even when not in order, earn 25 points. The sum of the points in each column added to the sum of the points in each row represents the grand total.

BINGO

Degree of difficulty: 1
Players: three or more

"Bingo" is very similar to "Lotto." The major difference is in the reward system. In "Lotto" the winner can brag about all the points that he has earned, but in "Bingo" the winner gets a prize. In "Bingo" each player pays a starting fee—let's say a piece of candy—and after the round, the winner gets all! A small fee is also paid to the umpire, because this game requires supervision.

The umpire has to prepare 99 "chips" from little pieces of paper, marked from "1" through "99". Every player (the more the merrier) has a card with a field of squares 3×4. These cards can be handled in two different ways: Either the umpire fills in the numbers he draws at random from the 99 pieces of paper, as in real "Bingo," or the players agree to do it themselves. Of course, each number can only be used once, since each number is only drawn once.

When the preparations are completed, the game can commence. A lucrative reward will generate a spirited game! The umpire begins to pull out one chip at a time and announces the number. It is a good practice to keep the chips in the order in which they were drawn in case of any dispute further

12	18	27	31
33	50	54	57
62	76	88	93

down the line. It is also important for the umpire to call out the number in a clear, loud voice so that everybody can hear it.

Players listen attentively to the announcer. Every number called that matches a number on a player's card must be crossed off. The player who crosses off all twelve numbers first calls "Bingo!" and gets a prize.

CONFLICT

Degree of difficulty: 6
Players: two

This game, which is based on topological principles, just like the game of "Sprouts," simulates the settlement of a new land. It is interesting to watch, during the course of the game, how two quarreling parties that occupy the same land rigidly defend themselves and must—eventually—separate from each other. It seems to be almost an organic process during which two equally strong parties can only survive if they have established a "neutral zone" between them.

In the process, every individual step has unforeseen and untold consequences. In theory, it's possible to anticipate certain outcomes in advance, but in the "field," or in practical terms, a player must use intuition, must roll with the punches, so to speak.

The tension of the conflict is highest when the game is played on a large piece of paper and the many different regions have been given rather high value points. For more "normal" games, however, a regular letter-size piece of paper is large enough. Both players need different colored pencils.

In the first phase of the game, the participants take turns drawing on an empty piece of paper one region at a time of any configuration. The spot for the first parcel of land can be anywhere as long as the subsequent plots are attached in such a way that the whole resembles a cluster of grapes (see the illustration). The players thus establish a country consisting of many different, connected regions. The players agree in advance on how many regions the country is to have; 20 or 30 is a useful number.

After the map has been laid out, both parties claim one region at a time in alternate turns. Both players start with an equal number of people they can settle on this land. Exactly how many should be determined in advance. As a guide, each side should have five or six times as many people as there are regions. This would make 100 people if the number of regions is 20.

Each player marks the region he claims in his respective color. On every turn, a player writes a number in an empty space. This number represents the number of people that are

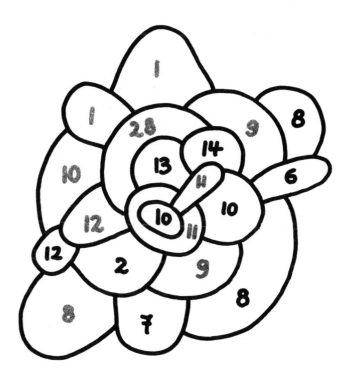

now living in that region. This continues until each party has settled all of its people. It is possible for a parcel to remain uninhabited.

No sooner have all the settlers staked out their claims than conflicts begin to arise. They are settled by the opponents taking turns. The first player attacks a neighbor, and he in turn fights back in like fashion.

At the beginning of each turn, a player states which one of the regions he intends to attack. The attacked region must defend itself against everyone of the attackers who border on its region. This means that the attacker is always in the majority and that the attacked region loses all of its settlers. The number in that region is crossed out, and the region is now defenseless.

The result of such actions is the creation of a zone of "dead" regions, which have become the dividing, or safety, zone between two quarreling factions. Eventually, the separation is complete, with none of the settled regions sharing a border and with the reason for the quarreling removed. The conflicts, in other words, have died out. Both parties now count the regions where their settlers are still present. Whoever has the largest number of regions claims victory. How many settlers have survived is of no consequence.

Cooperation

SCANDAL

Degree of difficulty: 2
Players: two or more

Scandals are created like folding pictures (see pages 121–124). The first person begins a story, passes it on to the second person, who expands on it and passes it on to the third person, who, again, adds some detail, until in the end, the story is revealed, without anybody knowing what anybody else was talking about.

Start by writing on a piece of paper the structure of the story:

1. the following people meet: female?
2. male?
3. where?
4. when?
5. what does she bring?
6. what does he bring?
7. what does she wear?
8. what does he wear?
9. what does she say to him?
10. what does he say to her?
11. what happens to both of them?
12. what do people say?

The first player writes down the answer to the first point on a piece of paper. He folds his portion of the paper back so that the next person doesn't know what he wrote. The paper is passed around the table in this manner, with each participant adding to the story. At the end, you unfold the paper and read the story. Here is an example of what can be expected:

Marilyn Monroe meets Ronald Reagan in an atom bunker on the day after graduation. She brings a chocolate Mickey Mouse and a toothbrush. He brings a remote-controlled intergalactic super rocket propelled by ions. She is wearing cotton pajamas. He wears a down suit. She says to him: "Dear, I want you to give me the stars from heaven." He says to her: "This day isn't over yet." Both are going to the movies together to see a Woody Allen film. People say it should never have gotten that far.

SUNDAY PAINTER

Degree of difficulty: 3
Players: two or more

The painter in this game is given a concept that he has to express in a drawing. The rest of the players watch and try to guess what the drawing describes. The player who guesses correctly takes the next turn.

Of course, the more original the subject or the concept is, the more fun you will have with this game. To draw, or to have guessed, a tree isn't really very exciting, but how about a concept like "the end of the workday"?

It's convenient to have a list of subjects and concepts ready before the game begins. Every participant can contribute by writing them on pieces of paper, which are then deposited in a container. The player, on his turn, picks one of them randomly and draws a picture according to the information on the paper.

FOLDING A PICTURE

Degree of difficulty: 3
Players: two or more

The joke with these folding pictures is that somebody finishes a drawing that somebody else started, without knowing what the first person drew. Viewing the sections together is worth the effort of the game.

The easiest version of "Folding a Picture" is the portrait. Fold a piece of paper lengthwise in half. On the left side of the

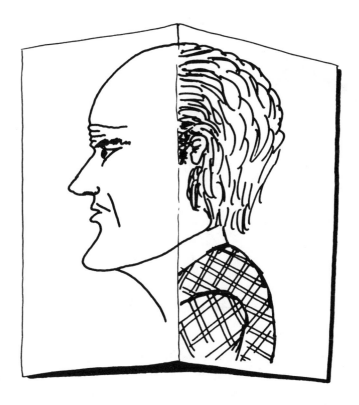

paper, the first artist draws the front portion of a head in profile: the forehead, nose, eyes, mouth, chin, and neck. On this side of the paper the artist can draw whatever he wants.

The drawing extends only slightly beyond the middle fold, with the right side remaining untouched.

The second artist receives the paper folded with the left side turned down so that he can't see what the first artist

drew. Only the few lines drawn beyond the middle fold give him something to build upon. The object now is to complete the portrait by drawing the back of the head on the right side of the paper. Of course, the head can be drawn full face with each player drawing one side of the face.

Another version of this game is where a whole person is drawn from head to toe—an ideal project when several people are present. Depending on the number of participants, more than one drawing can be done at the same time.

At the top of the paper, the first artist draws the head of a person down to the neck. The paper is folded back with only the beginning of the neck showing at the edge of the fold. The next artist is responsible for the torso, approximately down to the navel. The paper is folded again and handed to the next player, with only the edges of the previous drawing showing. This artist will draw from the navel to the knees; then he will fold the paper and hand it to the last player, who will complete the figure by drawing the legs and feet. The reward is watching the mutual effort unfold.

A very special version is the folding cartoon, which is the brainchild of Mario Grasso, a graphic artist in Switzerland. Fold a piece of paper so that part of the middle portion is hidden. Now draw a figure (or a picture) over that that extends over the fold and covers the left and right sides of the paper. The next step, which can be undertaken by another player, is to unfold the paper and fill in the blank portion in the middle. By opening and closing the fold, humorous or unexpected effects can be created in a continuously evolving drawing.

ARTWORK

Degree of difficulty: 3
Players: three or more

"Too many cooks spoil the broth!" Not so when it comes to drawing a picture. At least not in the game of "Artwork." If you can get your hands on a piece of paper, plus a few crayons, and if you can find a few friends with a little imagination, try this game!

The participants take turns adding a few details to a picture in progress—maybe only a dot, or a line, or some subtle shading. Before beginning the drawing, no discussion should take place about the motif that is being created. But every artist should have a certain picture in mind when he adds his detail to the drawing. But it's possible that if the previous player made a dot that he saw as a button on a uniform jacket worn by a knight on a horse that was about to jump through a fire, the next player might take this dot as being the eye of a dying swan on a moonlit, silvery lake.

The object and the fun of this game is to add such surprising, or even mischievous details to the drawing that the other players must constantly adjust and revise their concepts. The picture is completed when everybody agrees that no additional lines or strokes of a crayon can improve it.

Since a group of artists will have created this drawing, everybody should be represented when it comes to signing it. Let the participants, one by one, add letter after letter, and see what name evolves.

125

INDEX